MIX
Papier aus verantwortungsvollen Quellen
Paper from responsible sources
FSC® C105338

Abhishek Chauhan
Prof. Dr. P. Thakur

Power Quality Issues and their Impact on the Performance of Industrial Machines

Anchor Academic
Publishing

Chauhan, Abhishek, Thakur, P., Prof. Dr.: Power Quality Issues and their Impact on the Performance of Industrial Machines, Hamburg, Anchor Academic Publishing 2016

Buch-ISBN: 978-3-96067-081-0
PDF-eBook-ISBN: 978-3-96067-581-5
Druck/Herstellung: Anchor Academic Publishing, Hamburg, 2016

Bibliografische Information der Deutschen Nationalbibliothek:
Die Deutsche Nationalbibliothek verzeichnet diese Publikation in der Deutschen Nationalbibliografie; detaillierte bibliografische Daten sind im Internet über http://dnb.d-nb.de abrufbar.

Bibliographical Information of the German National Library:
The German National Library lists this publication in the German National Bibliography. Detailed bibliographic data can be found at: http://dnb.d-nb.de

All rights reserved. This publication may not be reproduced, stored in a retrieval system or transmitted, in any form or by any means, electronic, mechanical, photocopying, recording or otherwise, without the prior permission of the publishers.

Das Werk einschließlich aller seiner Teile ist urheberrechtlich geschützt. Jede Verwertung außerhalb der Grenzen des Urheberrechtsgesetzes ist ohne Zustimmung des Verlages unzulässig und strafbar. Dies gilt insbesondere für Vervielfältigungen, Übersetzungen, Mikroverfilmungen und die Einspeicherung und Bearbeitung in elektronischen Systemen.

Die Wiedergabe von Gebrauchsnamen, Handelsnamen, Warenbezeichnungen usw. in diesem Werk berechtigt auch ohne besondere Kennzeichnung nicht zu der Annahme, dass solche Namen im Sinne der Warenzeichen- und Markenschutz-Gesetzgebung als frei zu betrachten wären und daher von jedermann benutzt werden dürften.

Die Informationen in diesem Werk wurden mit Sorgfalt erarbeitet. Dennoch können Fehler nicht vollständig ausgeschlossen werden und die Diplomica Verlag GmbH, die Autoren oder Übersetzer übernehmen keine juristische Verantwortung oder irgendeine Haftung für evtl. verbliebene fehlerhafte Angaben und deren Folgen.

Alle Rechte vorbehalten

© Anchor Academic Publishing, Imprint der Diplomica Verlag GmbH
Hermannstal 119k, 22119 Hamburg
http://www.diplomica-verlag.de, Hamburg 2016
Printed in Germany

DEDICATED TO MY PARENTS

TABLE OF CONTENTS

CHAPTER 1
INTRODUCTION TO POWER QUALITY 5
1.1 Power Quality 5
1.2 Supply System 6
   1.2.1 Balanced Supply Voltage 6
   1.2.2 Unbalanced Supply Voltage 7
1.3 Causes of Unbalance Voltage 8
1.4 Interpretations of Voltage Unbalance 9
   1.4.1 National Equipment Manufacturer's Association (NEMA) 9
   1.4.2 IEEE Definition of Voltage Unbalance Analysis 9
   1.4.3 International Electrotechnical Commission (IEC) 10
   1.4.4 Non-standard Definitions 10
1.5 Effect of Unbalanced Voltage on the Performance of Three-Phase Induction Motor....11
   1.5.1 Derating of Three-Phase Induction Motor..12
   1.5.2 Inclusion of Under-Voltages and Over-Voltages 14
   1.5.3 Symmetrical Component Approach 14
   1.5.4 Importance of Angle of Unbalance on Induction Motor Performance...................17
1.6 Impact of Voltage Unbalance on Economy 18

CHAPTER 2
LITERATURE REVIEW AND MITIGATION TECHNIQUES 21
2.1 Literature of Assessment of Voltage Unbalance on Induction Motor 21
2.2 Mitigation Techniques for Unbalance Voltage 26
   2.2.1 Uniform Distribution of Single Phase Loads 26
   2.2.2 Proper Transposition of Transmission Lines..26

2.2.3	Line Conditioners	27
2.2.4	Static Voltage Ampere Reactive Compensator (SVC)	27
2.2.5	AC-Line and DC-Link Reactors to Adjustable Speed Drives	27
2.2.6	Special Purpose Transformer Configuration	28

CHAPTER 3
METHODOLOGY AND RESULT ANALYSIS 29

3.1	Analysis of Terminal Voltage and Problem Definition	29
3.2	Methodology	29
3.3	Assessment of Three-Phase Induction Motor with Speed Under Unbalanced Supply	33
3.3.1	Variation of Sequence Impedance with Slip	34
3.3.2	Variation of Efficiency and Input Power with Slip	35
3.3.3	Variation of Power Factor with Slip	37
3.3.4	Variation of Stator or Rotor Copper Losses with Slip	37

CHAPTER 4
CONCLUSION 39

REFERENCES 40

CHAPTER 1
INTRODUCTION TO POWER QUALITY

1.1 Power Quality

Power is the earliest incarnation, electricity is a kind of a magical force, something to be exhibited at the slideshow to curious awestruck onlookers. But it quickly became an essential part of a daily life, something now taken for granted by almost everyone in the industrialized world. At its most fundamental level, what it does give us light and heat when it is dark and cold. That is, electricity liberates humanity from the constraints of nature and contravenes the ordering of day and night. The power travels through poles and wires are an invisible yet vital force that connect us each to the other. But due to some acts the quality of this magical force is relegated. The worsening power quality (PQ) is the foremost concern when an emphasis is on the power system reforms. As per IEEE Std. 1159 [1], PQ refers to a wide variety of electromagnetic phenomena that characterize the voltage and current at a given time and location in the power system.

In simple terms, PQ is the combination of voltage and current quality, which is attributed mainly to the deviation of these quantities from the ideal, and is termed as PQ phenomena or disturbance [2]. Voltage disturbance is the most common form of a PQ problem in industrial utilities [3], [4]. Utilities are concentrating on utilizing the existing transmission systems more efficiently, users are focusing on high reliability and PQ. PQ is quantified technically on the basis of the constant sine-wave shape having no harmonics, constant frequency, symmetrical three-phase AC power system having equal magnitude of three voltages with phases shifted by 120°, constant root mean square value with nominal power system voltage value unchanged over time and fixed voltage in which the power system voltage unaffected by change in load [5].

A precise definition of the physically measurable disturbances and of limits specified on the basis of them is currently the subject of intensive work being carried out by standardization committees [6], [7]. Highly automated production processes are particularly susceptible to temporary changes in the magnitude and phase of the power supply voltage. Even voltage depressions lasting only a few milliseconds are enough to bring entire production lines to a standstill, causing considerable economic damage as well as endangering the production equipment itself. Sectors typically affected are the papermaking, semiconductor and chemical industries. Even when the power supply has been designed for maximum reliability, such disturbances cannot be completely ignored.

This report highlights the contribution of unbalance supply in the worsening of PQ and the effects on the three-phase induction machines (3-Φ IM) under the influence of unbalanced supply is also rigorously analyzed. Most widely used expression given by International Electrotechnical Commission (IEC) is used for the analysis of voltage unbalance.

1.2 Supply System

Large amount of power is generated at the generating stations, the present trend is to install bigger size of alternators to generate large amount of power to cater the required increasing demand. The site of the power of the generating station is depend on the type of generating station the new thermal station are being constructed at the pit heads because of higher cost of transportation of coal, whereas hydro power station sites are govern by availability of water resources [8]. The nuclear power plants are also situated remote from the centres of consumption due to safety reasons. Thus the difficulty of getting power station sites near the consuming sites makes it inevitable transfer bulk of electrical wires through longer distances. One of the major problem in supply system is to maintain the power as the balanced supply.

1.2.1 Balanced Supply Voltage

Three-phase voltages are said to be balanced if the magnitudes of the three voltages are equal and they are separated from each other by 120° (electrical) in phase as shown in Fig. 1.1. Because of the symmetry of the system and the balanced nature of the voltages, the analysis is made on a single-phase basis. This is due to that fact that a symmetrical system, because of the balanced voltages, gives rise to the balance current [6], [8].

$$\text{Red Phase: } V_{RN} = V_m \sin(wt) \tag{1.1}$$

$$\text{Yellow Phase: } V_{YN} = V_m \sin(wt - 120°) \tag{1.2}$$

$$\text{Blue Phase: } V_{BN} = V_m \sin(wt - 240°) \tag{1.3}$$

where,

V_m = Magnitude of voltage

V_{NR}, V_{NY} and V_{NB} are the line voltages of red, yellow and blue phases respectively.

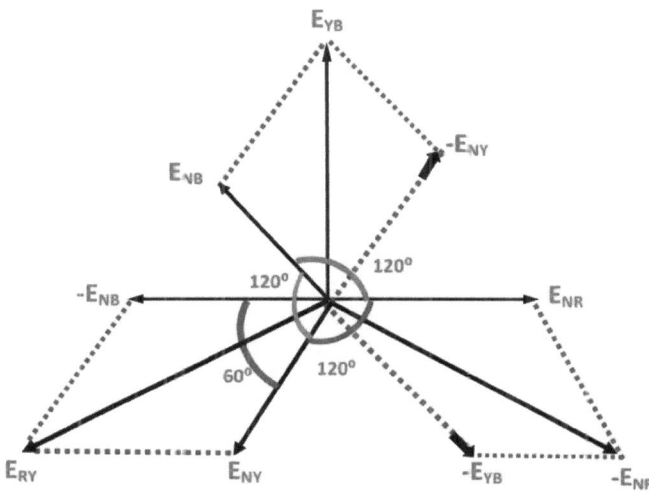

Fig. 1.1 Balanced voltage supply phasors [6]

1.2.2 Unbalanced Supply Voltage

With growing concern made by industries on PQ gives an incredible stimulus to the researchers for new researches in the direction of improving quality of power. Voltage unbalance is one of the major causes of downgrading of PQ which is observed in ample of industrial applications during last few decades.

Three-phase voltage supply is balanced in both magnitude and phase angle at generation and transmission levels, but the voltages at distribution end and consumption side it undergoes to unbalance [9]. Unequal voltage magnitudes at the fundamental frequency, fundamental phase angle deviation are the basic characteristics of voltage unbalance.

Voltage unbalance is regarded as any differences in the three phase voltage magnitudes or shift in the phase separation of the phases from 120 °. In three-phase balanced power systems the generated voltages are sinusoidal and equal in magnitude, with the individual phases 120 ° a part [6], [8], [9]. The nature of the unbalance includes unequal voltage magnitudes at the fundamental system frequency (under-voltages and over-voltages), fundamental phase angle deviation, and unequal levels of harmonic distortion between the phases. Voltage unbalance exists in almost all three-phase power system networks. The level of unbalance is considerably large in weak power systems and also those supplying large single phase loads [9].

7

Voltage unbalance describes the condition when the voltages of all phases of a 3-phase power supply are not equal. Practically, the three-phase supply is rarely balanced, according to American National Standards Institute report [10] only 66% of the 3-phase power delivered to industrial plants is within 1% voltage unbalance. Whereas, 98% of all voltage generated by electric utilities has 3% or less unbalance. Only 2% of the voltage produced by the electric utilities has a voltage unbalance greater than 3%.

1.3 Causes of Unbalance Voltage

The balancing problem becomes particularly difficult to compensate, when the unbalance is continually varying as with large industrial loads such as arc furnaces and adjustable speed drives. When a large number of single-phase adjustable speed drives (ASDs) are employed, this can result in continuously varying unbalanced loads [11]-[13]. ASDs are also nonlinear loads, with most topologies containing a diode rectifier front-end that draws very non-sinusoidal currents leading to harmonic distortion. The combination of ASDs, with the proliferation of single phase nonlinear switch-mode power supply based loads such as computers, can lead to unbalanced levels of distortion between phases which can also make the balancing process more challenging [11]-[13].

Other causes of unbalanced voltage supply are [9], [11]-[22]-:

(a) Uneven distribution of single-phase loads that can be continuously changing across a three-phase power system. For Example, rural electric power systems with long distribution lines, as well as large urban power systems where heavy single-phase demands, such as lighting loads, are imposed by large commercial facilities.

(b) Single-phase traction and electric transit and railroad systems can also cause considerable unbalance on the utility three-phase system unless proper design steps are taken.

(c) Incomplete transposition of transmission lines.

(d) Asymmetrical transformer winding impedances, open-wye (Y) and open-delta (Δ) transformer banks, asymmetrical transmission impedances possibly caused by incomplete transposition of transmission lines.

(e) Blown fuses on three-phase capacitor banks.

(f) The operation of non-linear loads like ASDs, power electronics equipment and frequent switching of single and three-phase loads are also accountable for supply voltage unbalance [13].

1.4 Interpretations of Voltage Unbalance

To quantify the severity of voltage unbalanced various standards have diverse definitions. In this section, different definitions of voltage unbalanced as prescribed in different standards are discussed to highlight their shortfalls.

1.4.1 National Equipment Manufacturer's Association (NEMA)

The severity of unbalance in supply voltage has been quantified by line voltage unbalance rate ($LVUR$) in NEMA and given as [23],

$$LVUR(\%) = \frac{\text{Max voltage deviation from avg line voltage}}{\text{Average line voltage}},$$

$$= \frac{Max\left[|U_{xy}-U_{av}|, |U_{yz}-U_{av}|, |U_{zx}-U_{av}|\right]}{U_{av}} \times 100\% \quad (1.4)$$

where $U_{av} = \dfrac{U_{xy}+U_{yz}+U_{zx}}{3}$

In this definition, only the magnitudes of line voltages (U_{xy}, U_{yz}, U_{zx}) have been considered to quantify the voltage unbalance. As line voltages are easy to measure in the field which makes this definition is well accepted by the industrial engineers. The major drawback of this definition is infinite sets of terminal voltages for fixed value of $LVUR$, each having a different influence on the performance of the motor [16]. Hence, NEMA definition is unsuitable to deliver preciseness to the assessment of 3-Φ IM operates under unbalanced supply [14], [15], [22] and [24].

1.4.2 IEEE Definition of Voltage Unbalance Analysis

The IEEE std. quantify the severity of unbalance in term of phase voltage unbalance rate ($PVUR$), which is given in [25] as

$$PVUR(\%) = \frac{\text{Max voltage deviation from avg phase voltage}}{\text{Average phase voltage}},$$

$$= \frac{Max\left[|U_{x}-U_{av}|, |U_{y}-U_{av}|, |U_{z}-U_{av}|\right]}{U_{av}} \times 100\% \quad (1.5)$$

where, $U_{av} = \dfrac{U_{x}+U_{y}+U_{z}}{3}$

The IEEE std. exercise similar definition for quantification of voltage unbalance as given by NEMA, the basic difference being that the IEEE uses the magnitude of phase voltages rather than line-to-line voltages. This definition is not found suitable for assessment of 3-Φ IM from field data as this definition is based on phase voltage. Further, the effect of angle unbalanced cannot be seen in this definition [17], [21].

1.4.3 International Electrotechnical Commission (IEC)

The degree of unbalance defined by IEC is named as a voltage unbalanced factor (*VUF*) also known as the true definition one compared with NEMA definition [15]. The value of *VUF* is defined as [26],

$$VUF(\%) = \frac{|U_n|}{|U_p|} \times 100\% = K_u \qquad (1.6)$$

where U_p and U_n are the positive and negative sequence component of the voltages in p.u.

VUF conveys improved information of the cause and effects of voltage unbalance on the motor as it directly reflects negative sequence component [21], [27]. But this definition alone is not found suitable for precise assessment of the performance of 3-Φ IM under supply voltage unbalanced as for fixed value *VUF*; there are infinite sets of terminal voltages. The precise prediction of the performance of the 3-Φ IM is only possible when conditions of under-or over-voltage is also available along with degree of voltage unbalance [22].

1.4.4 Non-standard Definitions

As in NEMA, IEEE and IEC definitions only the magnitude of unbalance is considered and significance of angle is entirely unnoticed. In [21], [27], a new complex quantity, known as complex voltage unbalance factor (*CVUF*) is introduced which specifies the angle along with the magnitude of *VUF*.

The angle of unbalance factor plays a vital role to choose the exact voltage set under the fixed degree of unbalance [21], [22], [27]. Moreover, peak losses and derating are stringently reliant on the angle of unbalance [12].

The *CVUF* is expressed as

$$CVUF = \frac{U_n}{U_p} = K_u \angle \theta_u \qquad (1.7)$$

where, θ_u is the angle of unbalance.

Field data generally do not have the information regarding the phase angle of line voltages which make the analysis with IEC definition is more challenging with field data. In [28], [29] an effort has been made to calculate the angle from the line voltages

To avoid the complex calculation associated with the symmetrical component approach, a fine estimate of the magnitude of *VUF* is explained in [15] as

$$\% \text{ Voltage Unbalance} = 82 * \frac{\sqrt{U_{xye}^2 + U_{yze}^2 + U_{zxe}^2}}{\text{average line voltage}} \quad (1.8)$$

where U_{xye}= difference between voltage U_{xy} and the average line voltage etc.

Angle calculated from line voltage is given in [28] as

$$\tan \theta_{te} = \frac{\sqrt{3}\left(U_{yz}^2 - U_{zx}^2\right)}{2U_{xy}^2 - U_{yz}^2 - U_{zx}^2} \quad (1.9)$$

By using (1.8), (1.9) precise analysis with the field data became feasible with IEC definition. In [22] it is asserted that efficiency, power factor, input power, rotor and stator copper losses are independent of angle of unbalance, hence, in this project IEC definition has been considered to put forward the precise prediction of performance of 3-Φ IM operating under unbalance supply.

1.5 Effect of Unbalanced Voltage on the Performance of Three-Phase Induction Motor

Ruggedness and relatively low operating cost make the 3-Φ IM most preferably used machine in the industrial applications. But the operation of 3-Φ IM are massively affected by the unbalance supply, moreover, unbalance supply results in the reduction in efficiency as shown in Fig.1.2, with increase in the intensity of unbalance supply the efficiency decreases, maximum percentage of VUF leads to the minimum efficiency. Additionally, unbalance supply leads to an increase in losses which further causes more power consumption and customer have to pay more for the same work [10], [14], [22], [29] and [30].

Moreover, unbalanced in supply possesses high winding temperature which causes much faster aging of the insulation system and hence, a significant shortening of machine operational life [31], [32]. In [18], the reduction in life of 3-Φ IM with rise in temperature has been estimated and it has been revealed that, the life of 3-Φ IM is reduced to half for every 10^0C rise in temperature. Aging of motor due to high winding temperature results in the replacement of induction machine before the actual time of replacement.

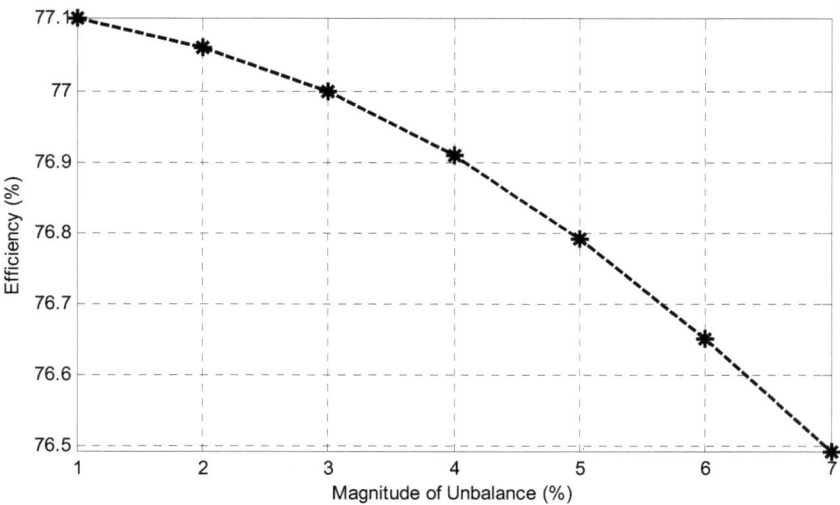

Fig. 1.2 Effect of unbalanced supply on the efficiency of three-phase induction motor [14]

Economic losses due to unscheduled downtime and premature shutdown of 3-Φ IM operating under unbalance supply, costs more than the expense associated with the replacement and reinstate of induction machine. Premature failure of the 3-Φ IM can only be prevented by derating of the machine which enables the safe operation of the machine within the thermal limits [15], [31]-[33].

1.5.1 Derating of Three-Phase Induction Motor

Performance of three-phase induction motor is explained on the basis of various parameters like efficiency, derating factor, power factor, input power, output torque, and total copper losses. Derating of motor is one of the most important factors which is to be strictly studied when the performance of induction motor under unbalance supply is analysed. Under full load conditions the motor is forced to operate at higher slip which causes the thermal losses in the windings and cause premature failure of the induction motor [22]. In order to prevent the motor from the condition of premature failure, simplest protection proposed by the NEMA standard, is to derate the motor to reduce its output horsepower load so it can tolerate the extra heating imposed by the unbalanced supply [31], [32], [33]. Derating of machine with different degree of unbalance is shown in Fig. 1.3, operating a motor for any length of time at voltage unbalance above 5% is not recommended. Derating factor is the ratio of mechanical output power under unbalance supply conditions to that under balanced conditions, where the machine is loaded in such a way the current do not exceed from the rated value [22], [34].

$$\text{Derating Factor} = \frac{\text{Mechanical Output Power Under Unbalanced Supply}}{\text{Mechanical Output Power Under Balanced Supply}} \qquad (1.10)$$

The NEMA standard states that once unbalance reaches 5%, the temperature begins to rise so fast that protection from damage becomes impractical [15], [17], [23].

There are several ways to develop a derating curve [15],

$$1 + \frac{\text{Increase in Winding Temperature Rise (\%)}}{100} = \left(\frac{\text{Load (\%)}}{100}\right)^{-1.7} \qquad (1.11)$$

$$1 + \frac{2(\text{Unbalance (\%)})^2}{100} = \left(\frac{\text{Load (\%)}}{100}\right)^{-1.7} \qquad (1.12)$$

The above relation can be used to find the percent load for operating under various percent unbalance.

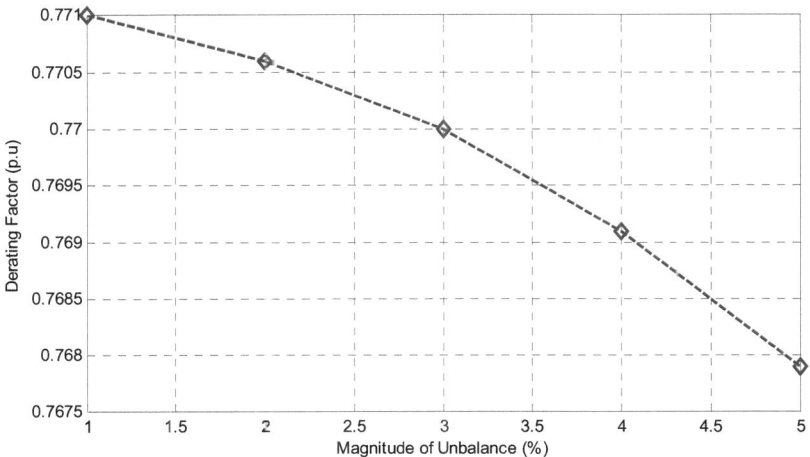

Fig 1.3 Derating for induction motor with degree of unbalance [23]

1.5.2 Inclusion of Under-Voltages and Over-Voltages

Studies performed in [15], [22], [31] recover that for the precise assessment of 3-Φ IM operates under unbalanced supply, condition of unbalance whether, under-or-overvoltage is to be considered along with the magnitude and angle of unbalance. In order to include over-voltages and under-voltages on the NEMA derating curve, the electrical and thermal models were developed in [15]. Fig. 1.4 shows the connection of the models. The electrical model is used to calculate motor losses. These losses are fed into the thermal model to predict the motor temperature rise. Then, the derating factor is calculated depending on the temperature magnitude.

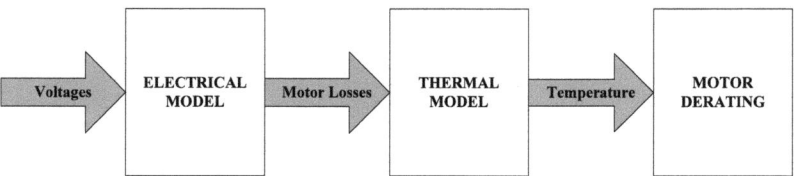

Fig. 1.4 Motor derating flow-chart [15]

1.5.3 Symmetrical Component Approach

Unbalanced system can be solved by a symmetrical per phase technique, which is known as the method of symmetrical components. C.L Fortescue proposed this method in 1918 [35]. Three unsymmetrical and unbalanced phasors of three-phase systems can be resolved into positive, negative and zero sequence component set of balanced phasors [35]. The effects of U_p and U_n on the performance of 3-Φ IM under unbalanced supply are shown in Fig 1.5. The effects of the zero sequence as the machine is ungrounded Y or delta Δ connected. A balance system would contain only positive sequence components of voltage, current and impedance. Therefore, the unbalanced motor voltage contains both U_p and U_n which have opposing phase sequences, i.e., "abc" and "acb", respectively [35].

As shown in Fig. 1.5, two opposing torques are generated by U_p and U_n, U_p produces the desired positive torque and responsible to describe the conditions of unbalance, whereas the negative sequence voltage produces an air gap flux rotating against the rotation of the rotor, which causes generating of an unwanted torque in reverse direction [10]. This result in the reduction of net torque and speed, and the possibility of speed, torque pulsations and increased motor noise [10], [15].

$$T_B > T_p - T_n \tag{1.13}$$

Where T_B = Torque under balance supply

T_p and T_n are the desired positive torque and reverse torque respectively

Moreover, the U_n in the unbalanced voltages generates large negative sequence currents due to the low negative sequence impedance, which increases the machine losses and temperatures [31]-[33]. Overall, overheating, line-current unbalance, derating, torque pulsation, inefficiency are the effects of unbalanced voltage on induction motor [9], [15]-[17], [31], [32]. C.Y. Lee [14] performed practical load test, on the basis of these tests it is investigated the impacts of unbalanced voltage supply on induction motor.

The analysis is rigorously focused on IEC definition as well as on the magnitude of the U_p and U_n. Eight conditions of unbalance 3ϕ-UV, 2ϕ-UV, 1ϕ-UV, 2ϕ-A, 1ϕ-A, 1ϕ-OV, 2ϕ-OV and 3ϕ-OV, on the foundation of these eight condition the effect on the efficiency and power factor with different magnitude of unbalance are illustrated. Fig.1.6 and Fig. 1.7 shows that, higher positive-sequence voltage gives a higher motor efficiency and a lower power factor respectively [14].

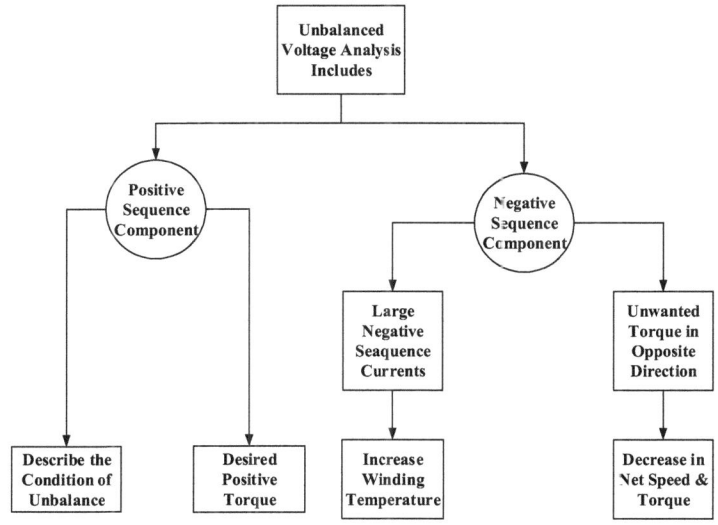

Fig. 1.5 Effect of symmetrical components on induction motor performance

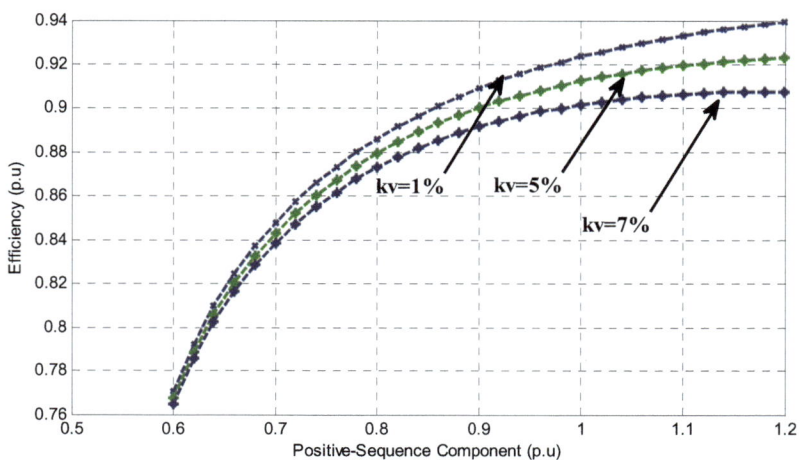

Fig. 1.6 Variation of efficiency with positive-sequence component under different kv (%) [14]

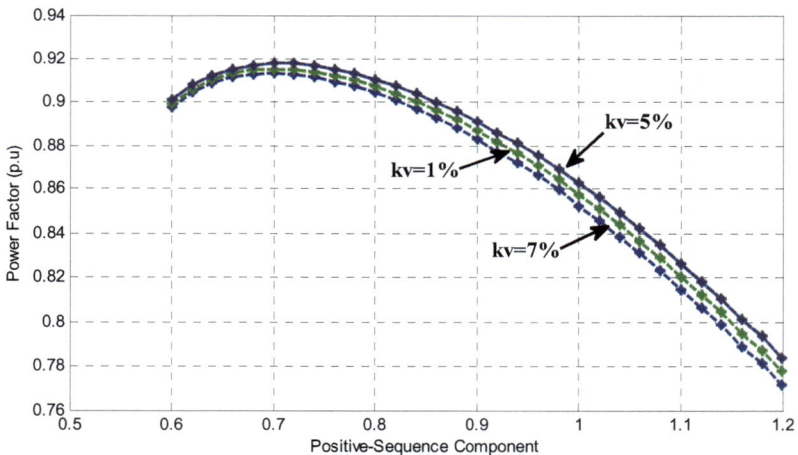

Fig. 1.7 Variation of power factor with positive-sequence component under different kv (%) [14]

The U_n also has contributions to the effects of unbalance on a motor's performance, U_n is responsible for the ill effects on the performance of induction motors under unbalance supply. Fig 1.8 revealed the contribution of U_n on efficiency and power factor, by putting U_p fixed and varies U_n such that VUF varied from 1% to 7%. It is observed that with an increase in U_n, efficiency decrease by 1.84% and has a little impact on the power factor i.e. decrease of about 0.48% [14].

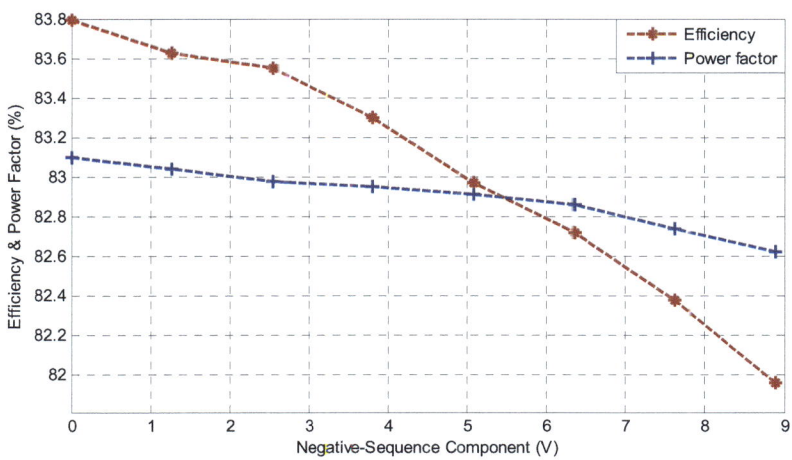

Fig. 1.8 Variation of efficiency and power factor with negative-sequence component under kv =1% to 7% [14]

1.5.4 Importance of Angle of Unbalance on Induction Motor Performance

A three phase supply have both the magnitude and the angle so in order to illustrate the actual impact of unbalance situations another expansion of *VUF* known as *CVUF* has been proposed by Y.J Wang in [21], [27] where phase angle has been considered in addition with IEC definition to counter the problem, this means that the *VUF* is extended as complex value under unbalanced supply differently due to the angles of the *CVUF*.

According to the classical approach, many of them are focusing only on the effect of the unbalance percentage on an induction machine, neglecting the influence of the angle between the U_p and U_n and a possible variation of the U_p. This approach was strongly criticized by Faiz *et al.* in [16]. They proved that neglecting the variation of the U_p and angle leads to great errors during the analysis of an asymmetrically fed induction machine.

Additionally, Kersting and Phillips in [34] pointed out that, it is not sufficient to merely know the percent voltage unbalance, but it is equally important to know how they are unbalanced, Y.J Wang [21] showed that the *CVUF* angle has a significant effect on currents, the allowable slip, and the derating factor of 3-Φ IM operates under unbalanced supply.

1.6 Impact of Voltage Unbalance on Economy

Due to various techno-economic advantages associated with induction motors, these are widely used in industrial, commercial and residential systems so any malfunctioning of motor may compel an effective financial trouble for power utilities, induction motor manufacturers, and consumers [12]. Unbalance voltage supply causes a depreciating impact on the efficiency of an induction motor [12], [14]-[18] decrease in efficiency cause higher electric motor usage for the same work which in-turn will have increasing noteworthy impact on electric bills. In Fig. 1.9 the economic losses due to unbalance in supply is highlighted.

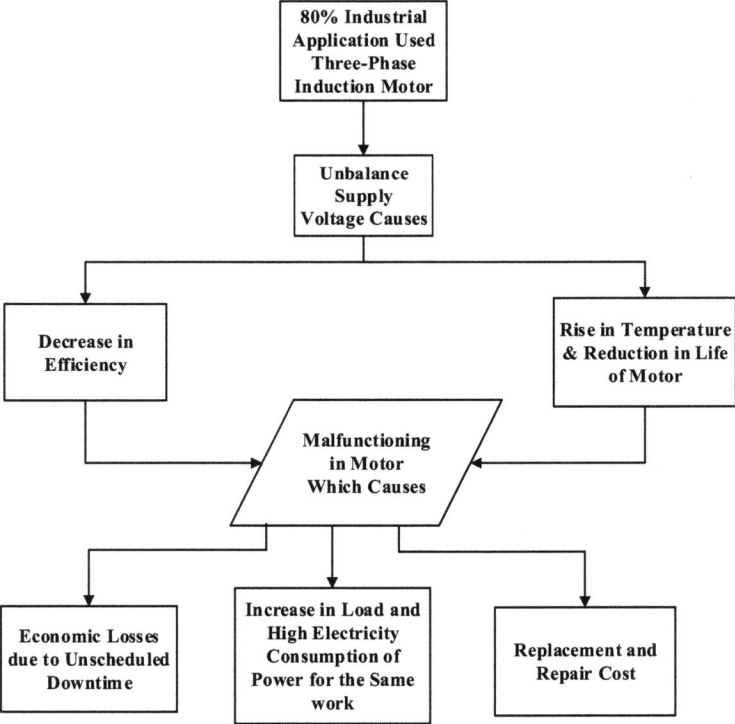

Fig. 1.9 Economic losses due to unbalance supply

Aging of motor due to high winding temperature results in the replacement of induction machine before the actual time of replacement. Economic losses due to unscheduled downtime and premature shutdown of 3-Φ IM operating under unbalance supply, costs more than the expense associated with the replacement and reinstate of induction machine.

Table 1.1 shown below explains an economic analysis performed by C.Y Lee in [14], on the basis of real load test analysis Lee calculate the extra consumption of power by a 1~5 HP 3-ø induction motor with *VUF* of 4% Lee assumed the average motor's efficiency under balanced rated voltage is 83.8% and average power consumption of 2.687kW. It is observe that under the unbalance of 4% a 1~5 hp 3- ø induction motor which operates an average of 2500 hours per year consumes an extra power of about 54174300kWh/yr and costs $3.14Million/yr.

Table 1.1 Power consumption of 1~5 hp 3ø induction motors when VUF is 4% [14]

Voltage Unbalance Cases	Total Installed Capacity(KW)	Motor Efficiency (%)	Load Increase Rate (LdIR)	Extra Power Consumption Per Year (KW/Yr)	Average Running Time Per Year (Hour)	Extra Electricity Consumption Per Year (KWh/Yr)	Extra Electricity Charge Per Year ($M/Yr)
Balanced	534000	83.8	1	0	2500	0	0
3ø-UV	534000	80.532	1.04058	21669.72	2500	54174300	3.142
2ø-UV	534000	81.382	1029701	15860.33	2500	39650825	2.307
1ø-UV	534000	81.505	1.023816	12717.74	2500	31794350	1.850
2ø-A	534000	82.254	1.018786	10031.72	2500	25079300	1.459
1ø-A	534000	83.041	1.009132	4876.48	2500	12191200	0.709
1ø-OV	534000	83.225	1.0069	3684.65	2500	9211625	0.536
2ø-OV	534000	83.402	1.004763	2543.44	2500	6358600	0.370
3ø-OV	534000	83.584	1.002578	1376.65	2500	3441625	0.2

For 3ø-UV:

Calculation for 3ø-UV is given as:

$$LIUU = LdIR - LIUB \qquad (1.14)$$

where,

LIUU= Load increase under unbalance condition

LdIR=Load increase rate

LIUB=load increase under balanced condition=1

- LIUU= 1.04058-1= 0.04058
- Extra power consumed/yr = (Total installed capacity) × (Increased load) = (534000) × (0.04058) = 21669.72kW/yr
- Extra electricity consumption/yr = (Extra power consumption/yr) × (Average running time/yr) = (21669.72 × 2500) = 54174300kWh/yr
- Extra consumption charges/yr paid by customer = (Extra electricity consumption/yr) × (Energy cost) = 54174300 × 0..058/kWh =$3.14Million/yr

Table 1.2 Impact of voltage unbalance on life span of induction motor winding [18]

Voltage Unbalance (%)	Winding Temperature (°C)	I^2R Losses (% in total)	Efficiency Reduction	Expected Winding Life (Years)
0	120	30%	Nil	20 years
1	130	33%	Up to ½%	10
2	140	35%	1-2%	5
3	150	38%	2-3%	2.5
4	160	40%	3-4%	1.25
5	180	45%	5% or more	Less than 1

In Table 1.2 it is revealed that how poorly the unbalance voltage decreases the lifespan of a motor and the cost is again bear by the motor operator. I^2R Losses in the rotor and stator, causes the motor to run hotter, for every 10° C increase in winding temperature insulation life is cut by half [18]. The decrease in life span due to increase in temperature of an induction motor when operates in unbalance supply also been methodically discussed in [31] by using

$$\text{Temperature of Winding} = 2 \times (\% \text{ of Unbalance})^2 \tag{1.15}$$

Table 1.3 Loss to the industries due to shut down of plant under unbalanced condition [18]

Industry	Cost of Down Time, $/hour
Paper & pulp	15,000
Petro-chemical	150,000

In Table 1.3 cited approx estimation of loss due to replacement, repair costs for premature motor failures and the accurate cost of unbalanced voltage. The cost of motor associated with an industrial application is very less in comparison with the loss faced by the industry during unscheduled downtime and low production rate. Electric Power Research Institute estimated that industries collectively lose $45.7 billion a year due to outages and in addition lose $6.7 billion each year to PQ phenomena in which unbalanced voltage is an important concern which is to be strictly observed [37].

CHAPTER 2
LITERATURE REVIEW AND MITIGATION TECHNIQUES

2.1 Literature of Assessment of Voltage Unbalance on Induction Motor

Assessment of voltage unbalance is not a new focus of research, different machines are effected differently under unbalanced supply. It will take about six decades of research to find out diverse and appreciable factors, who precisely describe the performance of 3-Φ IM machines operates under unbalanced supply.

J.W William in 1954 studied about the operation of three-phase induction motors on unbalanced voltages [38]. W.C Gafford *et. al.* in 1959 elucidate excess temperature rise due to the unbalanced voltage, which shortens the life span of an induction motor [39]. A new method of calculating reduced motor rating due to voltage unbalance is projected by N.L Schmitz and M.M Berndt in 1963 [33]. R.F Woll in 1975 explain the fact that a large manufacturing plant may have well balanced incoming supply voltage, but unbalance can develop within the plant from its own single-phase power requirements if these are not uniformly spread among the three phases [40].

Table 2.1 Literature based on the Analysis of three-phase induction motor under unbalance supply

References	Findings
[38]	Reduction in efficiency.
[39]	Premature motor failure due to heating.
[33]	Derating of poly-phase induction motors
[40]	Most of time non-uniform distribution of single phase loads in an industry results the unbalance in its own supply line.
[41]	Determine operating quantities of squirrel cage IM as a function of unbalanced voltage and results of this study is used for selection of relays to provide protection against voltage unbalance.

These aforementioned studies quoted in Table 2.1 shows only present qualitative results. The definition of unbalanced voltage and the resulting motor characteristics were not reflected in these studies [16]. Further, various important issues, as for example, the negative effect of negative and positive sequence component of voltage on different performance parameters of 3-Φ IM, the effects of overvoltage etc, have been ignored in these study [14], [30].

Furthermore, the effects of deviation of phase angle associated with supply voltage unbalance on the performance of the 3-Φ IM have been ignored in these studies. The various parameters of 3-Φ IM are very sensitive to this angle. As for example, the peak stator and rotor current, peak loss, derating factor etc are dependent of angle of unbalance [21], [22]. Further, for the same percentage of unbalance, the pattern of three phase voltage also depends on the deviation of phase voltage angle associated with voltage unbalance and its inclusion for the precise assessment of the performance of 3-Φ IM is required.

Additionally, contributions made toward the assessment of the performance of 3-Φ IM under the condition of unbalance have been summarized in Table 2.2. These studies considered either NEMA standard or IEC definition for the degree of unbalance. These studies give more precise information then the studies presented in Table 2.1, as in most of these studies *VUF* has been considered to quantify the degree of unbalance. *VUF* gives the true value of voltage unbalance as this index is very close to negative sequence component of the voltage and this component of voltage is the primary cause of voltage unbalance [21].

Table 2.2 Literature of assessment of induction motor performance under supply voltage unbalance based on various definitions and factors

Ref.	Definitions Considered	Findings & Shortfalls
[14]	VUF U_p and U_n	**Findings-:** For same *VUF*, $U_p \alpha \dfrac{(\eta)^x}{(2f)^y}$ and $U_n \alpha \dfrac{1}{\eta^z}$ **Shortfall-:** Only limited conditions of unbalance have been discussed. • The angle of unbalance factor has been ignored.
[21]	VUF & Angle of unbalance	**Findings-:** A new factor *CVUF* is introduced. • Magnitude as well as angle of *CVUF* is considered. • Same magnitude of *CVUF* may affect the operation of an IM differently due to different angles of *CVUF*. **Shortfall-:** Computational errors are still high. • Analysis only limited to MATLAB and Simulation environment. • Not effectively used for field analysis because angle is not present in field data.
[15]	NEMA	**Findings-:** NEMA derating curve of induction machines has been extended to include over-voltages and under-voltages. **Shortfalls-:** Other performance parameters of IM in under and over-voltage conditions are not explained.

[16]	$LVUR$, VUF, & Angle of unbalance factor.	**Findings-:** NEMA, VUF and $CVUF$ are not complete. • NEMA definition is not complete because there are several assumptions used in previous papers shows that average terminal voltage of the machine is assumed equal to the rated voltage. • It is hard to predict the exact derating for particular value of unbalance until the average values of voltage is correctly identified. • VUF only is not complete, for a particular VUF (%) there are infinite combinations of terminal voltages which affect the performance of 3-Φ IM accordingly. **Shortfall-:** Analysis limited to MATLAB environment.
[17]	VUF & Angle of unbalance factor.	**Findings-:** To avoid wide range of derating factor in the derating process of IM, a method is suggested which give exclusive numerical value of derating factor for any degree of unbalance. **Shortfall-:** There is still a large range of derating factor. • For a particular VUF (%) there is a complete range of derating factor for different angles.
[22]	VUF, Angle of unbalance factor & Coefficient of unbalance (f).	**Findings-:** For precise assessment coefficient of unbalance (f) is inserted along with the magnitude and angle of $CVUF$. • 'f' shows the condition of voltage deviation. **Shortfalls-:** Impact of negative sequence components with $CVUF$ is ignored. • Magnetic loss, mechanical loses etc is ignored. • This study is only based on simulation. • This assessment method cannot be used with field data.

Although the degree of unbalance defined by IEC express better physical understanding of the negative effect of supply voltage unbalance on the performance of 3-Φ IM, but this index suffers from the following limitations:

- Quantify the degree of unbalance only, the conditions of unbalance do not reflect in this method.

- The angle of unbalance factor plays vital role to decide the exact voltage set under the fixed degree of the unbalance. In [21], it has been revealed that, for the fixed degree of unbalance, there are infinite numbers of voltage sets.

- The need of the magnitude as well as angle of three phase voltages in calculation of *VUF* makes this definition unsuitable for the assessment of the performance of the 3-Φ IM with field data, as field data generally do not consists the information regarding the phase angle of phase or line voltages. The remedy of this problem is competently explained in [28] where angle of unbalance is determined from the magnitude of the line voltages. Due to this approach IEC definition becomes more proficient for the analysis with the field data.

- Further, the complex algebra involves in symmetrical component analysis makes the calculation of *VUF* complicated.

To resolve the anomalies exist in the method of quantification of degree of unbalance, the definition of unbalance factor has been extended in [21] by incorporating the angle of unbalance factor. The new index which is defined in [21] is named as *CVUF* which comprises both the magnitude as well as the angle of unbalance factor. But in this study the variation of positive sequence component of voltage has not been considered in the assessment of the performance of 3-Φ IM under the condition of supply voltage unbalance.

In [16], it has been proved that, neglecting the variation of positive sequence component of voltage leads to a great error in the performance analysis of the 3-Φ IM under the condition of supply voltage unbalance. The same discussion has been presented in [14] and [22]. In [22], both the angle and the variation of positive sequence component in terms coefficient of unbalance have been considered to precise the assessment of the performance of 3-Φ IM under the condition of supply voltage unbalance. But this study is based on simulation only. Further, the magnetic loss, mechanical loss etc. have not been considered in this study and hence this method of assessment is far from the real assessment.

2.2 Mitigation Techniques for Unbalance Voltage

In order to maintain balance in the supply system and restrain down the ill impacts on economy numerous mitigation techniques have been suggested in literature, most of them are efficiently used by different power utilities to confront the problem of unbalanced voltage.

While several mitigation techniques have been suggested to correct voltage unbalance, maintaining an exact voltage balance on all three phases at the point of use is virtually impossible for the following reasons [42]-:

a) Single-phase loads are continually connected to, and disconnected from, the power system.
b) Single-phase loads are not evenly distributed between the three phases.
c) Power systems may be inherently asymmetrical.

Therefore, some voltage unbalance will be present in any type of low-voltage three-phase three wire or three-phase four-wire systems [43].

2.2.1 Uniform Distribution of Single Phase Loads

Uneven distribution of single phase loads across the three phases is the basic cause of supply unbalance, by equal distribution of single-phase loads across all three phases help to minimize the problem of unbalance voltage up to an extent [10]. A balance distribution system can be obtained by changing the configuration of system through automatic and manual feeder switching operations to relocate load with circuits [44].

2.2.2 Proper Transposition of Transmission Lines

Transposition is the episodic swapping of positions of the conductors of a transmission line. The unbalance of the line, which can lead to one-sided loads in three-phase systems, is also condensed by complete transposition [8]. Whereas improper transposition causes the unbalance of voltage, but in long transmission lines multiple transpositions become unfeasible when there is an increase in substations connected to transmission lines. Additionally, transposition of conductors amplify puzzlement when addressing urgent fault situations, due to this unbalance introduced by the incomplete transposition is not comparatively noteworthy, modern power lines are not often transposed [8].

2.2.3 Line Conditioners

An active line conditioner is proposed in [45] to maintain balance in a three phase AC system. The proposed system publicized that the injection of a correction voltage in any one phase is adequate to cancel the negative sequence voltage component in the incoming three-phase supply and the resulting voltage obtained at the load terminal are fundamentally positive sequence voltages and therefore are balance. Due to termination of the negative sequence voltage component by the proposed scheme significantly improves the performance of induction motor connected to an unbalanced AC system [45].

2.2.4 Static Voltage Ampere Reactive Compensator (SVC)

Special fast-acting power electronic circuits, such as SVC can be configured to limit the unbalance [46], [47]. A. Campos et al. [47] projected a system where shunt connected thyristor-controlled SVC is used for voltage unbalance in AC supply having variable loads associated with it. Through a three-phase transformer a three-phase pulse width modulation voltage source inverter connected in series with the line which is responsible for nullification of negative-sequence voltage. The basic shortcoming observed in this approach is of slow response, injection of harmonic into the AC system and the necessity for large passive components which in turn increase the cost of the system.

With a newly develop individual phase control scheme a fixed capacitor-thyristor controlled reactor (FC-TCR) type of SVC can be applied in the distributing system, an FC-TCR type SVC can trim down negative sequence current caused by the uneven distribution of single-phase loads to recover the balance condition in the distribution network [46]. Microcomputer associated with the control circuit, replaces the traditional discrete load switching and enables the capability of fast and active balancing of the system. Concurrently an appropriate selection of inductor and capacitor values may responsible for improved power factor.

2.2.5 AC-Line and DC-Link Reactors to Adjustable Speed Drives

The ASDs as energy-saving schemes are employed in various industrial applications for speed control process in drives which cause a continuous variation of load, deployment of great number of single-phase ASDs results in continuously varying unbalanced loads and make balancing process more critical and challenging. Some test results were specified in [43] and suggested that connecting both AC-line and DC-link reactors to the ASDs has greatest effect on phase-current unbalance and reduced it by near to half [13].

An LC filter in the drive's input rectifier stage can also be used to attenuate these undesired effects of unbalance voltage [13]. Relay protection system related methods is also employed for the protection of induction motors from unbalanced voltages [41] and reliability of negative sequence current relays explained in [48].

2.2.6 Special Purpose Transformer Configuration

As discussed in chapter 1 rapid intensification of electric railways and single-phase traction systems also responsible for the unbalance of voltage across the three phase line. Some transformers like Scott and Steinmetz used as an important tool for the improvement of unbalanced supply so that the application can run under considerable unbalance conditions [49].

In some emergent situations, a three-phase induction motor with Steinmetz connection can run with single-phase supply, at the price of considerable unbalance in its line voltages and currents [20]. A Steinmetz-transformer is in fact a three-phase transformer with an extra power balancing load, consisting of a capacitor and an inductor rated proportional to the single phase load.

It is essential to sustain balance between the power electronics based systems which are dynamically shown its significance for performing various control processes in numerous industrial application and the unbalance in supply cause by them, since a proper modesty among them facilitates the efficient performance of all the equipment allied with the distribution system through balance supply.

CHAPTER 3
METHODOLOGY AND RESULT ANALYSIS

3.1 Analysis of Terminal Voltage and Problem Definition

The terminal voltage variations for fixed degree of unbalanced has been properly discussed in [14], [21], [22] and [30]. In these studies, it has been revealed that, for fixed degree of voltage unbalance, there are infinite sets of terminal voltages, and for each set of terminal voltages the behavior of induction motor is unique. Hence, the precise assessment of the performance of induction motor is only possible when information about the set of terminal voltages is available.

Further, to reduce the range of terminal voltage variations, inclusion of U_p, angle of unbalance (θ_u) and condition of unbalance (f) are suggested in [14], [21], and [22], respectively. All of these aforementioned studies do not consider the effects of voltage unbalance on the speed of 3-Φ IM, in this project it is affirmed that,

- the speed of 3-Φ IM gives the reflection of conditions of unbalance and independent from degree of unbalance,
- for a fixed condition of under-or over-voltage, the slip is independent from degree of voltage unbalance,
- the other performance parameters of 3-Φ IM, viz, efficiency, power factor, copper loss, etc, do not depend only on the degree of unbalanced but precise and unique assessment of these parameters are possible only if information about the conditions of voltage unbalance are also available. Speed of the 3-Φ IM gives the information of conditions of unbalance, hence its inclusion along with the degree of unbalanced for precise the assessment of performance of 3-Φ IM has been asserted.

Due to these external attributes of slip, this project incorporates the values of slip under different conditions of supply voltage unbalance and various performance parameters have been analyzed.

3.2 Methodology

Proposed methodology is showcased in Fig. 3.1, various performance parameters have been analyzed with different values of slip under different conditions of supply voltage unbalance. It has been publicized that, performance parameters of 3-Φ IM can be assessed precisely if the information of slip is available along with the magnitude of unbalanced.

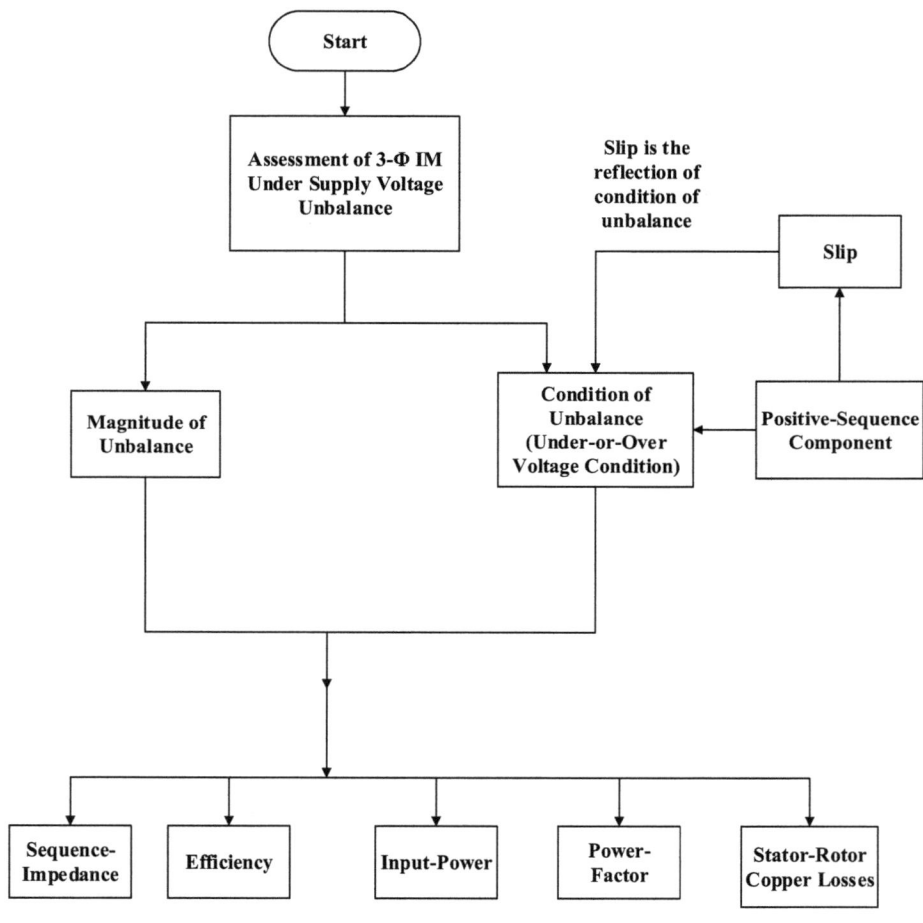

Fig. 3.1 Block diagram of methodology used

Positive and negative sequence equivalent circuits of 3-Φ IM are shown in Fig. 3.2. Equivalent circuit has R_s and R_r as the stator and rotor resistance, X_s and X_r are the stator and rotor reactance, I_{sp} and I_{sn} represent stator positive and negative sequence current, I_{rp} and I_{rn} represent the rotor positive and negative-sequence currents and U_{sp} and U_{sn} are the positive and negative sequence component of voltage for stator.

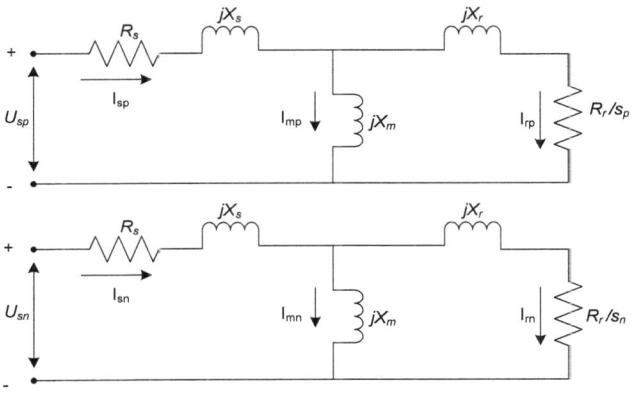

Fig 3.2 Sequence equivalent circuit of three-phase induction motor

It is assumed that all the circuit elements are constant, mechanical and stray losses are neglected, and the motor is Δ-connected. Due to Δ-connection there is no zero-sequence component U_{sp} and U_{sn} runs the motor with slip s_p and s_n respectively.

$$s_p = s \tag{3.1}$$

$$s_n = 2 - s \tag{3.2}$$

where s_p and s_n is for forward and backward slip respectively in p.u.
slip can be calculated by using

$$slip(s) = \frac{N_s - N_r}{N_s} \tag{3.3}$$

where N_s and N_r are synchronous speed and rotor speed respectively in rpm.

The positive sequence input impedance (Z_p) and negative-sequence input impedance (Z_n) are obtained from Fig. 3.2 as,

$$Z_p = R_s + jX_s + \frac{(jX_m)\left[\left(R_r/s_p\right) + jX_r\right]}{\left(R_r/s_p\right) + j(X_m + X_r)} \tag{3.4}$$

$$Z_n = R_s + jX_s + \frac{(jX_m)\left[\left(R_r/S_n\right) + jX_r\right]}{\left(R_r/S_n\right) + j(X_m + X_r)} \tag{3.5}$$

Positive and negative sequence stator current is given by (3.6) and (3.7), respectively

$$I_{sp} = \frac{U_{sp}}{Z_p} \tag{3.6}$$

$$I_{sn} = \frac{U_{sn}}{Z_n} \tag{3.7}$$

Motor input active power is given as

$$P_i = \text{Re}\left|3\left[U_{sp}(I_{sp}^2) + U_{sn}(I_{sn}^2)\right]\right| \tag{3.8}$$

and input reactive power is expressed as

$$Q_i = j\left|3\left[U_{sp}(I_{sp}^2) + U_{sn}(I_{sn}^2)\right]\right| \tag{3.9}$$

Motor active output power is given as

$$P_o = \left[3|I_{pr}^2|\left[\frac{1-S_p}{S_p}\right] \times R_r\right] + \left[3|I_{nr}^2|\left[\frac{1-S_n}{S_n}\right] \times R_r\right] \tag{3.10}$$

Motor efficiency is explained as

$$\eta = \frac{P_o}{P_i} \times 100\% \tag{3.11}$$

The power factor can be estimated as follows

$$pf = \cos\left[\tan^{-1}\left(Q_i/P_i\right)\right] \tag{3.12}$$

where Q_i and P_i are the motor active and reactive power, respectively.

3.3 Assessment of Three-Phase Induction Motor with Speed Under Unbalanced Supply

In this section the performance of 3-Φ IM with the speed under different conditions of unbalance has been rigorously discussed, the variations of slip with the positive sequence component of voltage for different values of *VUF* have been investigated and shown in Fig. 3.3. The ratings of 3-Φ IMs have been considered as,

- Motor-I ratings [21], [22]: 240 V (line-to-line), 7.5 kW, 60 Hz, six poles, Δ-connected. The resistance and reactance of its equivalent circuit in ohms per phase referred to the stator and rotor are: R_s=0.294 Ω, X_s= 0.503 Ω, R_r= 0.144 Ω, X_r=0.209 Ω, X_m=13.25Ω.

- Motor-II ratings [50]: 460 V (line-to-line), 18.64 kW, 60 Hz, six poles, Δ-connected. The resistance and reactance of its equivalent circuit in ohms per phase referred to the stator and rotor are: R_s=0.641Ω, X_s= 1.106 Ω, R_r= 0.332 Ω, X_r=0.464 Ω, X_m=26.3 Ω.

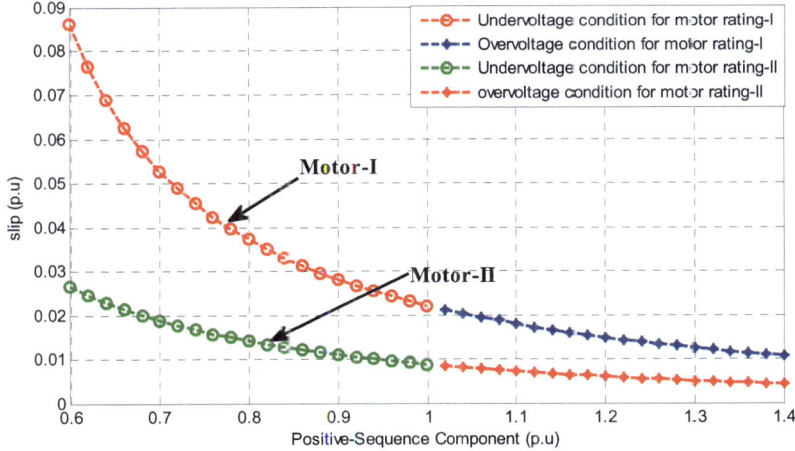

Fig. 3.3 Change in slip with positive- sequence component of voltage for k_v equals to 3% and 5%

From Fig. 3.3 the following observations have been extracted,

- The values of slip are only dependent on the values positive-sequence component of voltages. It is independent from *VUF* if the value of U_p is maintained constant. It can be seen from Fig. 3.3 that, for Motor I, the value of slip is 0.05 p.u for k_v equals to 3% and 5%, both, if U_p equal to 0.7. Similarly, for Motor II, the value of slip is 0.02 under similar conditions. Hence, it can be said that, the slip is *VUF* independent parameter and only depends on positive sequence component of voltages, which gives the reflection of condition of unbalance according to [14].

- In under-voltage condition, where $U_p<1$, slip is more than the values of slip in over-voltage conditions, where $U_p>1$, for any value of *VUF*. Hence, small values of slip represent over-voltage conditions whereas large values of slip represent under-voltage conditions. The ranges of slip in over-and under-voltage conditions depend on the rating of the 3-Φ IM. For example, ranges of slip in under-voltage conditions, for Motor-I, are 0.02-0.09, whereas same, for Motor II, are 0.01-0.03, respectively. Hence with the knowledge of slip, it is easy to know about the conditions of unbalanced.

- To reduce the terminal voltage variations, for fixed value of *VUF*, the knowledge of conditions of unbalance is very essential [14], [22] and hence slip can be used as a parameter which can be used to represent the conditions of unbalance.

3.3.1 Variation of Sequence Impedance with Slip

The exceeding of stator current above the rated value results in overheating of stator windings [21], [22] and [31], as from eq. (3.6) and (3.7) it is scrutinized that I_{sp} and I_{sn} are the functions of respective sequence voltages and impedances. Further, from eq. (3.4) and (3.5), it seems that Z_p and Z_n are the function of slip, in order to prevent the overheating of stator windings the slip must be limited to a certain value [21], for this the variation of impedance with slip is required to be analyzed under different conditions and degree of unbalance.

Results are analyzed by using motor-I, in Fig. 3.4, it is highlighted that,

- sequence impedances are independent of degree of unbalance,

- Z_n remains constant throughout the range of slip and under various conditions of unbalance. Hence, it is asserted that under supply voltage unbalance, I_{sn} is influenced by the change in U_{sn} only,

- whereas, Z_p have a nonlinear relationship with the slip, over- or under-voltage condition respectively possesses maximum and minimum of Z_p. Hence, from eq. (3.6), U_p and Z_p are mutually accountable to affect the I_{sp},

- if sequence impedances on a particular slip is achieved, than estimation of sequence voltages became feasible to maintain the stator currents below the rated values.

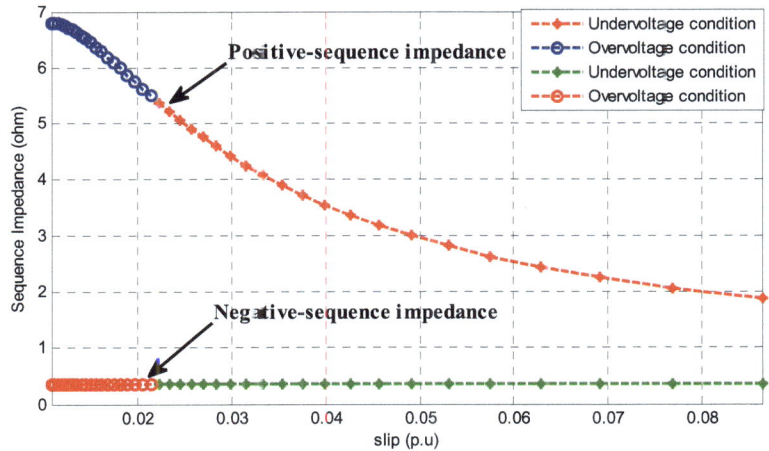

Fig. 3.4 Variation of positive-sequence input impedance with change in slip for k_v equals to 3% and 5%

3.3.2 Variation of Efficiency and Input Power with Slip

The downgrading in efficiency is reported in various literatures, when operation of 3-Φ IM is under unbalanced supply [10], [21] and [22], hence, inclusion of preciseness in the assessment of efficiency gain tremendous popularity. Variation of efficiency with slip under different condition of unbalance is shown in Fig. 3.5, and it is revealed that,

- efficiency is degree of unbalance dependent parameter, particular degree of unbalance have a complete range of efficiency, conditions of unbalance and slip is used to draw out the exact value of efficiency. The motor operating with under-voltage unbalance and higher slip will experience lower efficiency than if it operates with over-voltage unbalance conditions and lower slip.

35

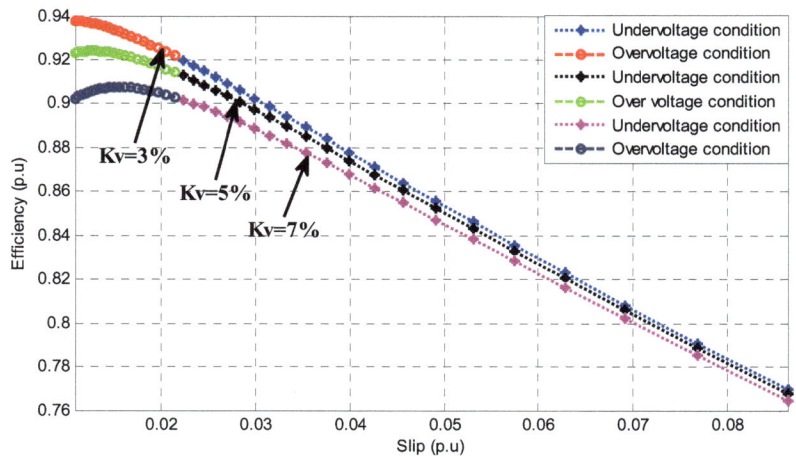

Fig. 3.5 Variation of efficiency with change in slip

- It has been discussed, from Fig. 3.5 and Fig. 3.6, that efficiency reduction in under-voltage high slip region leads to an increase in total power consumption, and customers have to bear an increased consumption expense.

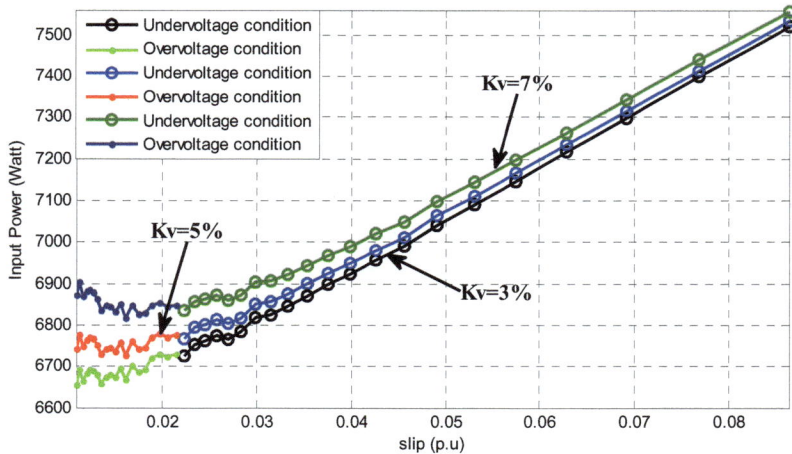

Fig. 3.6 Variation of input power with change in slip

3.3.3 Variation of Power Factor with Slip

The variation of power factor with change in slip under different degree of unbalance is shown in Fig. 3.7. It is studied that power factor is slightly influenced by degree of unbalance, whereas, under-or-over voltage conditions have shown a significant variation in power factor. On analyzing Fig. 3.5 and Fig. 3.7, it is examine that lower efficiency possesses higher power factor in under-voltage high slip region, when compared with over-voltage low-slip region.

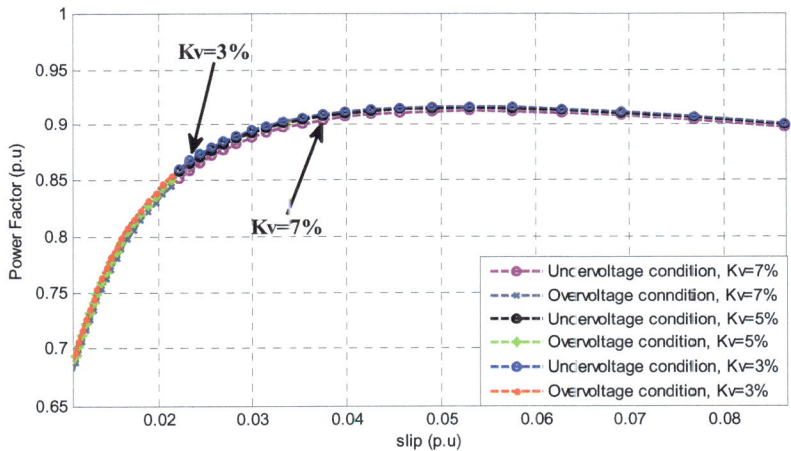

Fig. 3.7 Variation of power factor with change in slip

3.3.4 Variation of Stator or Rotor Copper Losses with Slip

Change in the stator (P_s) and rotor copper losses (P_r) under unbalance supply with slip are analyzed in Fig. 3.8 and Fig. 3.9 respectively and studied that,

- the variation of P_s and P_r are quite similar, copper losses are degree and conditions of unbalance dependent parameters,
- losses are more prominent at under-voltage high slip region, whereas, lower slip results in low copper losses.

Analysis shown in Fig. 3.5 to Fig. 3.9 explains the performance of 3-Φ IM with change in speed, which in turn affect the slip of 3-Φ IM operates under unbalance supply. Importance of speed and conditions of unbalance is highlighted to precise the assessment of performance parameters.

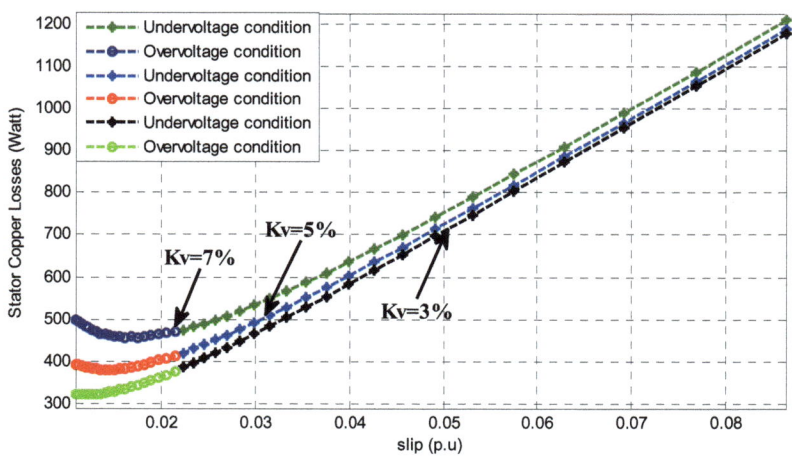

Fig. 3.8 Variation of stator copper loss with change in slip

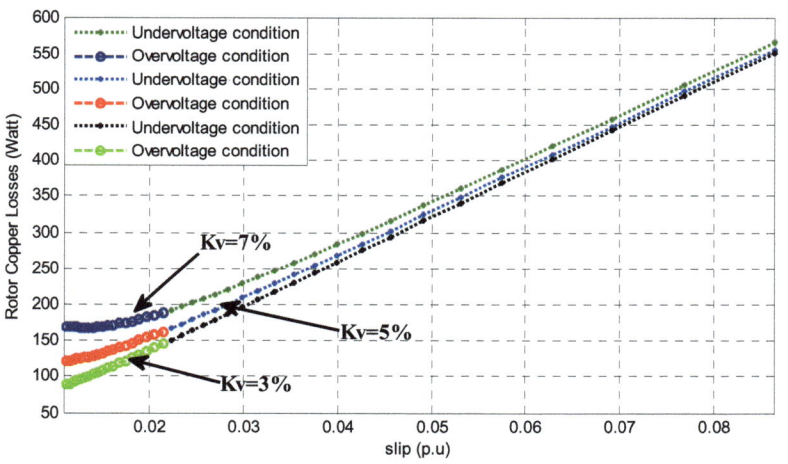

Fig. 3.9 Variation of rotor copper loss with change in slip

CHAPTER 4
CONCLUSION

The degree of voltage unbalance is a quantity required in studying the performance of a three-phase induction motor operating with an unbalanced voltage supply. This project highlights the impact of unbalance supply on the performance of 3-Φ IM. Degree of unbalance, condition of unbalance and slip are all simultaneously studied to analyze the performance characteristics of 3-Φ IM under unbalance supply. Efficiency, power factor, input power and stator-rotor copper losses are strictly affected by the magnitude of unbalance, so IEC definition is enough to explain the performance of 3-Φ IM. It has scrutinized that U_p have a tendency to influence the input impedance and slip, and independent of the degree of unbalance. Efficiency is the degree of unbalance dependent parameter, higher degree of unbalance possesses low efficiency in under-voltage condition with higher slip Whether, low efficiency results in higher power factor, higher stator-rotor copper losses and increase in rate of power consumption.

It is also concluded that condition of unbalance is a motor dependent parameter and it change with the change in the ratings of the motor. This analysis enables the operator to know the performance of 3-Φ IM at different speeds of the motor when operate under unbalanced supply. The tuning of the machine according to these characteristics results the operation of 3-Φ IM under recommended thermal limits and minimize the ill impacts on its performance causes due to unbalance supply. Additionally, some mitigation techniques which are commonly used for minimizing the severity of unbalance supply are also discussed where SVR gain comprehensive popularity in modern times.

REFERENCES

[1] *IEEE Recommended Practice for Monitoring Electric PQ*, IEEE Std. 1159-1995, 1995.

[2] M. H. J. Bollen, *"Understanding PQ Problems"* NewYork: IEEE Press, 2000.

[3] D. Ezer, R. A. Hanna, and J. Penny, "Active voltage correction for industrial plants," *IEEE Trans. Ind. Appl.*, vol. 38, no. 6, pp. 1641–1644, Nov. 2002.

[4] A. Elnady and M. M. A. Salama, "Mitigation of voltage disturbance using adaptive perception-based control algorithm," *IEEE Trans. Power Del.*, vol. 20, no. 1, pp. 309–318, Jan. 2005.

[5] T. Jauch, A. Kara, M. Rahmani and D. Westermann, "Power quality ensured by dynamic voltage correction", *ABB High Voltage Technologies Review*, pp-25-36, Apr. 1998.

[6] R. C. Dugan, M. F. Mc Granaghan and H. W. Beatys *"Electrical Power Systems Quality"*. McGraw-Hill, New York, Madrid, Montreal, Sydney Tokyo, 1996.

[7] *IEEE Recommended Practices and Requirements for Harmonic Control in Electrical Power Systems*, IEEE Std 519–1992, Jan. 1993.

[8] J. J. Grainger and W. D. Stevenson, *"Power System Analysis"*, McGraw- Hill, Inc., 1994.

[9] A.V Jouanne and B. Banerjee, "Assessment of Voltage Unbalance," *IEEE Trans. Power Del.*, vol. 16, no. 4, pp. 782-790, Oct. 2001.

[10] *Electric Power Systems and Equipment-Voltage Ratings (60 Hertz)*, ANSI Standard Publication no. ANSI C84.1-1995.

[11] C. Y. Lee, B. W. Chin, W. J. Lee, and Y. F. Hsu, "Effects of various unbalanced voltages on the operation performance of an induction motor under the same voltage unbalance factor condition," in *Proc. IEEE Ind. Commercial Power Syst. Conf.*, pp. 51–59, 1997.

[12] C.Y Lee, B.K Chen, W.J Lee, Y.F Hsu, Effects of various unbalanced voltages on the operation performance of an induction motor under the same voltage unbalance factor condition, *"ELSEVIER: Electric Power Systems Research"* pp.153-163, 1998.

[13] K. Lee, G.Venkataramanan and T.M. Jahns, "Modeling effects of voltage unbalances in industrial distribution systems with adjustable-speed drives", *IEEE Trans. on Ind. Appl.*, vol. 44, no. 5, Sept. 2008.

[14] C. Y. Lee, "Effects of unbalanced voltage on the operation performance of a three-phase induction motor," *IEEE Trans. Energy Convers.*, vol. 14, no. 2, pp. 202–208, Jun. 1999.

[15] P. Pillay, P. Hofmann, and M. Manyage, "Derating of induction motors operating with a combination of unbalanced voltages and over or undervoltages," *IEEE Trans. Energy Convers.*, vol. 17, no. 4, pp. 485–491, Dec. 2002.

[16] J. Faiz, H. Ebrahimpour, and P. Pillay, "Influence of unbalanced voltage on the steady state performance of a three-phase squirrel cage induction motor," *IEEE Trans. Energy Convers.*, vol. 19, no. 4, pp. 657–662, Dec. 2004.

[17] Jawad Faiz and H. Ebrahimpour, "Precise Derating of Three-phase Induction Motors with Unbalanced Voltages", *in proc. IEEE Ind. Appl.*, pp. 485-491, Jun. 2005.

[18] "Energy matters industrial technology program: The effects of unbalanced voltage on the life and efficiency of three-phase electric motors", *U.S Department of Energy and Renewable Energy*, Mar. 2005.

[19] R. Bergeron, "Voltage unbalance on distribution systems phase I," *Canadian Electrical Association, Montreal, Quebec*, vol. 231, pp. 488-494, Jan. 1989.

[20] T. A. Kneschke, "Control of utility system unbalance caused by single phase electric traction," *IEEE Trans. Ind. Appl.*, vol. 21, no. 6, pp. 1559–1570, Nov. 1985.

[21] Y. J. Wang, "Analysis of effects of three-phase voltage unbalance on induction motors with emphasis on the angle of the complex voltage unbalance factor," *IEEE Trans. Energy Convers.*, vol. 16, no. 3, pp. 270–275, Sept. 2001.

[22] Makbul Anwari and A. Hiendro, "New unbalance factor for estimating performance of a three-phase induction motor with under and overvoltage unbalance", *IEEE Trans. Energy Convers.*, vol. 25, no. 3, Sept. 2010.

[23] *Motors and Generators*, ANSI/NEMA Standard MG1-1993.

[24] P.Giridhar, R.C Bansal, and R.S Aithal, "A novel approach towards interpretation and application of voltage unbalance factor", *IEEE Trans. Ind. Electronics*, vol. 54, no. 4, Aug. 2007.

[25] *IEEE Standard Test Procedure for Polyphase Induction Motors and Generators*, IEEE Standard 112, 1991.

[26] *Testing and Measurement Techniques - Unbalance, Immunity Test*, IEC Standard 61000-4-27, Aug. 2000.

[27] Y.J. Wang, "An Analytical Study on Steady-State Performance of an Induction Motor Connected to Unbalanced Three-Phase Voltage", *in proc. IEEE Power Engineering Society Winter Meeting, Singapore*, pp. 159-164, Mar. 2000.

[28] S.G.Jeong, "Representing Line Voltage Unbalance,", *37th IAS annual conference meeting, Pittsburgh, PA, USA*, vol. 3, pp 1724-1732, Oct. 2002.

[29] S.B Singh, A. K. Singh and P. Thakur, "Assessment of induction motor performance under voltage unbalance condition", in *Proc. Harmonics and Quality of Power conf.*, Hong Kong, pp-256-261, Jun. 2012.

[30] C.Y Lee, B.K Chen, W.J Lee and Y.F Hsu, "Effects of various unbalanced voltages on the operation performance of an induction motor under the same voltage unbalance factor condition", *Electric Power Syst. Research* vol. 47, no. 3, pp. 153-163, Nov. 1998.

[31] P. Gnacinski, "Windings temperature and loss of life of an induction machine under voltage unbalance combined with over- or under-voltages", *IEEE Trans. Energy Convers.*, vol. 23, no. 2, Jun. 2008.

[32] P. Pillay and M. Manyage, "Loss of life in induction machines operating with unbalanced supplies", *IEEE Trans. Energy convers.*, vol. 21, no. 4, Dec. 2006.

[33] N. L. Schmitz and M. M. Berndt, "Derating poly-phase induction motors operated with unbalanced line voltages," *IEEE Trans. Power Appl. Syst.*, pp. 680–686, Feb. 1963.

[34] H. Kersting, W.H Phillips, " Phase Frame Analysis of the Effects of Voltage Unbalance on Induction Machines", *IEEE Trans. on Ind. Appl.*, vol. 33, pp. 415-420, 1997.

[35] C. L. Fortescue, "Method of symmetrical coordinates applied to the solution of polyphase networks," *AIEE Trans.*, vol. 37, pp. 1027–1140, 1918.

[36] D. S. Dorr, M. B. Hughes, T. M. Gruzs, R. E. Jurewicz, and J. L. Mc-Claine, "Interpreting recent PQ surveys to define the electrical environment," *IEEE Trans. Ind. Appl.*, vol. 33, no. 6, pp. 1480–1487, Nov. 1997.

[37] EPRI, *The Cost of Power Disturbances to Industrial and Digital Economy Companies*. Palo Alto, CA,. EPRI Executive Summary, 2001.

[38] J. W. Williams, "Operation of 3 phase induction motors on unbalanced voltages," *AIEE Trans. Power Appl., Syst.*, vol. PAS-73, pp. 125–133, Apr. 1954.

[39] W. C Gafford, Duesterhoeft and C. C. Mosher, "Heating of Induction Motors on Unbalanced Voltages," *AIEE Trans. Power Apparatus and Syst.*, vol. 78, pp. 282-297, Jun. 1959.

[40] R. F. Woll, "Effect of unbalanced voltage on the operation of polyphase induction motors," *IEEE Trans. Ind. Appl.*, vol. 11, no. 1, pp. 38–42, Jan. 1975.

[41] P. B. Cummins, J. R. Dunki-Jacobs, and R. H. Kerr, "Protection of induction motors against unbalanced voltage operation," *IEEE Trans. Ind. Appl.*, vol. 21, no. 4, pp. 778–792, May 1985.

[42] D. S. Dorr, M. B. Hughes, T. M. Gruzs, R. E. Jurewicz, and J. L. Mc-Claine, "Interpreting recent PQ surveys to define the electrical environment," *IEEE Trans. Ind. Appl.*, vol. 33, no. 6, pp. 1480-1487, Nov. 1997.

[43] C. N. Souto, C. de Oliveira and M. Neb, "Induction motor thermal behaviour and life expectancy under non-ideal supply conditions", *IEEE Trans. on Energy Convers.*, vol.21, no.6, May 2000.

[44] R. P. Broadwater, A. H. Khan, H. E. Shaalan, and R. E. Lee. "Time varying load analysis to reduce distribution losses through reconfiguration," *IEEE Trans. Power Del.*, vol. 8, no. 1, pp. 294–300, Jan.1993.

[45] V. B. Bhavaraju and P. N. Erjeti, "An active line conditioner to balance voltages in a three-phase system," *IEEE Trans. Ind. Appl.*, vol. 32, no. 2, pp. 287-292, Mar. 1996.

[46] J.H Chen et al, "Using a static VAR compensator to balance a distribution system", *in proc. of IEEE Ind. Appl. Conf.*, vol.4, pp. 2321 - 2326, 1996.

[47] A. Campos, G. Joos, P. D. Ziogas, and J. F. Lindsay, "Analysis and design of a series voltage unbalance compensator based on a three-phase VSI operating with unbalanced switching functions," *IEEE Trans. Power Electronics*, vol. 9. no. 3, pp. 259–274, May 1994.

[48] P. Lerley, "Applying unbalance detection relays with motor loads," *IEEE Trans. Ind. Appl.*, Vol. 35, no. 2, pp. 689–693, May 1999.

[49] J. Driesen and T.V Craenenbroeck, "Voltage disturbances introduction to unbalance", *Copper Development Association IEE Endorsed Provider*, May 2002.

[50] T. Wildi, *"Electric Machines, Drives and Power Systems"*, 6th Edition, PEARSON: 2005.